Walter Hill

Collection of Queensland Timbers

Walter Hill

Collection of Queensland Timbers

ISBN/EAN: 9783744642255

Printed in Europe, USA, Canada, Australia, Japan

Cover: Foto ©berggeist007 / pixelio.de

More available books at **www.hansebooks.com**

Printed for Distribution at the Melbourne International Exhibition of 1880.

BOTANIC GARDENS, BRISBANE.

COLLECTION

OF

QUEENSLAND TIMBERS,

COLLECTED AND ARRANGED BY

WALTER HILL,

Colonial Botanist, and Director of the Botanic Gardens, Brisbane.

THE several collections of Queensland Timbers which have been exhibited at the International Exhibitions at London, Paris, Philadelphia, and Vienna, have attracted much attention on account of their variety and value for commercial purposes. The knowledge that they are adapted for all the uses to which timber is applied, and that special qualities are possessed by some of the trees represented in this collection, is rapidly extending, with, it is trusted, beneficial results to the colony.

Many varieties of the Eucalyptus have found their way to different parts of the globe, and the demand for seed from those countries which desire to introduce this genus is in excess of what Queensland is able to supply. In India and on the Continent of Europe the naturalization of the tree has been successful. The uses to which the several species can be applied are enumerated in the following list of exhibits.

There are still extensive scrubs of the Red Cedar (*Cedrela Toona*) in the colony, especially in the districts north of Cardwell. Of late years there has been a wholesale destruction of this valuable timber; and in view of the supply of cedar failing from this cause, the Government have introduced a Bill into Parliament, having for its object the conservance of this timber and that in the State forests generally.

CONIFERÆ.

1. *Araucaria Bidwilli*, Hook. (Bunya Bunya).—Diameter, 30 to 58 inches; height, 100 to 200 feet. A noble tree, inhabiting the scrubs in the district between Brisbane and the Burnett Rivers. In the 20th parallel it grows thickly over a portion of country, in extent about 30 miles long and by 12 broad. The tree has a very singular appearance; the trunk is quite straight; its bark is thick and smooth; the branches are produced in whorls of six, seven, or eight; they are horizontal, inflexed, and ascending at the extremities. From the style of growth, singular foliage, and peculiar fresh colour, when surrounded with other trees of a different habit and greyish tint, it produces a fine effect, from the striking contrast presented by its rigid growth, and fresh green lance-shaped leaves. The wood is not only very strong and good, but it is full of beautiful veins, and capable of being polished and worked with the greatest facility. The cones produced on the extreme upper branches, with their apex downwards, are large, measuring 9 to 12 inches in length, and 10 inches in diameter; on coming to maturity they readily shed their seeds, which are 2 to 2½ inches long by 1 inch broad, sweet before being perfectly ripe, and after that resemble roasted chestnuts in taste. In accordance with regulations issued by the Government, the tree is not allowed to be cut down by those who are licensed to fall timber on the Crown lands, the fruit being used as food by the aboriginals. The trees produce some cones every year, but the principal harvest happens only every three years, when the blacks assemble from all quarters to feast on it. The food seems to have a fattening effect upon them, and they eat large quantities of it, after roasting it at a fire. Contrary to their usual habits, they sometimes store up the Bunya nuts, hiding them in a water-hole for a month or two. Here they germinate, and become offensive in taste to a white man's palate, but are considered by the blacks to have then acquired an improved flavour. The taste of the Bunya when fresh has been described as something between a chestnut and a raw potato.

2. *Araucaria Cunninghamii*, Ait. (Moreton Bay Pine).—Diameter, 36 to 60 inches; height, 150 to 200 feet. This majestic tree is, without exception, the most ornamental and useful tree in Queensland. Its beautiful regular pyramidal form, and the sombre green of its awl-shaped foliage, command general admiration. It covers immense tracts of land along the coast, and in the interior. It overtops all other trees, whether growing on the alluvial banks near rivers, or upon the steep and rugged mountains in the interior. Its branches are produced in whorls from six to eight, horizontally and spreading. The bark is thick and brownish. The timber is an article of great commercial importance, and is used extensively in this colony. The wood is strong and durable when kept dry, but soon decays when exposed to alternate damp and dry. When procured from the moun-

tains in the interior it is fine-grained, and is susceptible of a high polish, which excels that of satin-wood or bird's-eye maple. The resin which exudes is very remarkable; it has all the transparency and whiteness of crystal, and that portion of it which adheres to the trees hangs from them in the shape of icicles, which are are sometimes 3 feet long, and 6 to 12 inches broad. The sawyers receive at the present time 6s. 6d. to 7s. per hundred superficial feet, some trees yielding as much as 10,000 feet of saleable timber.

3. *Dammara robusta*, C. Moore. (Kauri or Dundathu Pine).—Diameter, 36 to 72 in.; height, 80 to 130 feet. This huge tree inhabits the alluvial banks on the rivers near the coast in the Wide Bay District ; also in the moist and sheltered valleys on Frazer's Island. It has a smooth-barked trunk, of a red colour ; the branches are produced in whorls from 5 to 10, distant, spreading, and of a large size. The wood is fine-grained, free of knots, and easily worked. It is, however, not a plentiful tree. At the present time the sawyers are receiving 7s. 6d. per hundred superficial feet ; some trees yield as much as 25,000 feet.

4. *Frenela Endlicheri*, Parlat. (Cypress Pine).—Diameter, 20 to 40 inches; height, 50 to 70 feet. This tree forms vast tracts along the coast, growing on barren and sandy soils. The timber is an article of great importance ; it is durable, fine-grained, fragrant, and capable of a high polish ; it is used for piles of wharves and for sheathing punts and boats; it resists the attacks of cobra and white ants, and the root is valued by cabinet-makers for veneering purposes. The market value at the present time is 10s. per hundred superficial feet.

5. *Frenela rhomboidea*, Endl. (Cypress Pine).—Diameter, 12 to 18 inches ; height, 40 to 50 feet. A handsome tree, scattered through the brigalow scrubs and sand ridges on the Darling Downs District. The timber is much used for telegraph posts, and by settlers for building purposes.

6. *Frenela Parlatorei*, F. M. (Mountain Cypress Pine).—Diameter, 12 to 24 inches; height, 40 to 60 feet ; timber much valued for cabinet purposes.

7. *Podocarpus elata*, R. Br. (She-pine).—Diameter, 20 to 36 inches ; height, 50 to 80 feet. A very beautiful tree, trunk rarely cylindrical, timber free from knots, soft, close, easily worked ; good for joiners' work, and used for spars. It occurs very frequently in the scrubs along the coast. The market value at the present time is about 7s. per hundred feet superficial.

CASUARINEÆ.

8. *Casuarina equisetifolia*, Forst. (Swamp Oak).—Diameter, 12 to 24 inches; height, 50 to 70 feet. Found growing in great abundance near salt-water marshes and inlets. The wood is coarse-grained, and beautifully marked, it is used for purposes where lightness and toughness are required.

9. *Casuarina torulosa*, Ait. (Forest Oak or Beefwood).—Diameter, 9 to 18 inches; height, 30 to 40 feet. A small tree occupying large tracts of land in the open forest. The timber is much used for fuel ; it is close and prettily marked, and gives handsome veneers.

10. *Casuarina glauca*, Sieb. (The River She-Oak).—Diameter, 24 to 36 inches ; height, 70 to 90 feet. A robust tree of general occurrence

on the borders of rivers and creeks. The timber is strong and tough, used for staves, shingles, &c.

11. *Casuarina Cunninghamiana*, Miq. (Scrub She-Oak).—Diameter, 6 to 12 inches; height, 20 to 36 feet. A small but handsome tree; timber hard, close, and prettily marked.

MELIACEÆ.

12. *Cedrela toona*, Roxb. (Red Cedar).—Diameter, 24 to 75 inches; height, 100 to 150 feet. This magnificent deciduous tree is found in scrubs along the coast, and occasionally extending inland for a considerable distance. It puts out large branches, the junctions of which with the stem furnish those beautiful curled pieces of which the choicest veneers are made. The timber is light, very durable, easily worked, and is largely employed in house-joinery and furniture-making; in fact, whenever lightness and durability are required. It is an article of great commercial importance, and is largely exported to the other colonies. The market value at the present time is from 20s. per hundred superficial feet, according to colour and size.

13. *Flindersia australis*, R. Br. (Flindosa).—Diameter, 36 to 48 inches; height, 80 to 100 feet. A robust tree of general occurrence in the scrubs on the banks of rivers. The timber is hard, close, and of great strength and durability, and would make excellent timber for railway purposes. It shrinks very little in drying, and does not rust iron. It has long been known to timber merchants as being a very hard timber and difficult to cut with the saw, and for that reason little attention has been paid to procuring it.

14. *Flindersia Oxleyana*, F. M. (Light Yellow-wood).—Diameter, 24 to 42 inches; height, 80 to 100 feet. This fine tree is found in the same situations as red cedar. The timber is strong, durable, fine-grained, and of good colour, used in boat-building, cabinet-work, and for many of the purposes to which cedar is applied. It also possesses dyeing properties. The present market value is from 7s. to 9s. per hundred superficial feet.

15. *Flindersia maculosa*, F. M. (Spotted Tree of the Colonists).—Diameter, 12 to 18 inches; height, 36 to 40 feet. A middle-sized tree; the trunk spotted by the falling off of the outer bark in patches. Timber used for shingles and staves of tallow casks and pick handles.

16. *Flindersia Bennettiana*, F. M. (Bogum Bogum).—Diameter, 18 to 26 inches; height, 70 to 90 feet. This large tree occurs in most of the coast scrubs; the timber is close-grained, but is seldom used. It splits well; might, probably, be valuable for staves.

17. *Flindersia Schottiana*, F. M.—Diameter, 18 to 30 inches; height, 30 to 60 feet. This moderate-sized tree extends from the boundary of New South Wales to the Daintree River. It is confined generally to forest gullies. The timber is hard, close-grained, and prettily marked.

18. *Owenia venosa*, F. M. (Sour Plum).—Diameter, 9 to 36 inches; height, 30 to 40 feet. A moderate-sized tree, common in the brigalow scrubs. Its great strength renders it suitable for many purposes.

19. *Owenia acidula*, F.M. (Rancooran.)—Diameter, 12 to 18 inches; height, 30 to 50 feet. A handsome small tree, occurring in the brigalow scrubs of West Moreton and the Darling Downs; wood close-grained, and nicely marked.

20. *Amoora nitidula*, Benth.—Diameter, 18 to 30 inches; height, 70 to 90 feet. A large-sized tree of frequent occurrence in scrubs bordering the coast. Qualities of the timber not much known.

21. *Synoum glandulosum*, A. Juss.—Diameter, 18 to 24 inches : height, 40 to 60 feet. A moderate-sized tree of very general occurrence ; timber firm and easily worked.

22. *Dysoxylon Muelleri*, Benth. (Pencil Cedar).—Diameter, 20 to 40 inches; height, 70 to 80 feet. A large-sized tree inhabiting the rich alluvial scrubs upon the banks of the rivers in the districts of Moreton Bay and Wide Bay. Timber of a rich colour, used for cabinet-making and window-work. When fresh cut the timber has much the smell of a Swedish turnip. Market value, from 6s. to 6s. 6d. per hundred feet.

23. *Dysoxylon rufum*, Benth. (Bastard Cedar Pencil Wood).—Diameter, 18 to 24 inches ; height, 40 to 50 feet. A slender tree, occurring in many of the scrubs on the coast, and also in the interior. Wood is nicely grained, and used for various purposes, but principally for cabinet-work.

23A. *Dysoxylon Fraseranum*, Benth. (New South Wales Rosewood).—Diameter, 24 to 36 inches; height, 80 to 100 feet. Of frequent occurrence in the scrubs about Highfields. Wood fragrant ; used for various purposes, but principally for cabinet-work, for turnery and for carving.

24. *Melia composita*, Willd. (White Cedar).—Diameter, 15 to 24 inches; height, 40 to 50 feet. An elegant deciduous tree, never ranging very far from the coast. Timber soft and easily worked ; not in very good repute, though undeservedly, as the timber from a well-matured tree is found to be very durable.

PALMÆ.

25. *Livistona australis*, Mart. (Cabbage-tree).—Diameter, 12 to 18 inches ; height, 40 to 80 feet. The lofty stem is covered with leaves in a dense mass, orbicular in circumscription when fully out, 3 to 4 feet diameter. The young leaves can be plaited as a material for hats.

26. *Livistona inermis*, R. Br. (Partridge Wood).—Diameter, 12 to 15 inches ; height, 14 to 40 feet. A beautiful and handsome tree : the outer portion of its trunk is very hard, beautifully marked, and takes a good polish.

27. *Ptychosperma elegans*, Blume. (Bangalow).—Diameter, 6 to 12 inches ; height, 60 to 80 feet. A lofty magnificent feathery-leaved palm-tree ; its wood is used for rails for fences. It is destined to take here a prominent position in decorative plantations.

28. *Ptychosperma laccospadix*, Benth. (Black Palm).—Diameter, 6 to 8 inches ; height, 12 to 16 feet. A very handsome palm. The outer portion of the wood is used for making walking-sticks.

PANDANE.E.

29. *Pandanus pedunculatus*, R. Br. (Screw Pine).—Diameter, 6 to 12 inches ; height, 18 to 30 feet. This singular and stately plant for scenic group planting occurs in great abundance on the coast lands. The fruit is used by the aborigines.

FILICES OR FERNS.

30. *Alsophila australis*, R. Br. (Tree Fern).—Diameter, 3 to 9 inches: height, 20 to 50 feet. This beautiful tree is frequent in shaded ravines and permanently damp scrubs ; wood prettily marked.

RUTACEÆ.

31. *Bosistoa sapindiformis*, F.M.—Diameter, 6 to 12 inches; height 15 to 20 feet. A small but very handsome tree, abounding in most of the scrubs near the sources of the Logan and Albert Rivers. Wood close and light.

32. *Citrus australis*, Planch. (Native Orange.)—Diameter, 6 to 14 inches ; height, 20 to 30 feet. This small and handsome tree grows in abundance on the borders of scrubs, both on the coast and in the interior. The trunk is erect, with many diffused branches, armed with axillary straight thorns of about half an inch long. The fruit is about one and a-half inches in diameter, almost globular; an agreeable beverage is produced from its acid juice. The wood is hard, close-grained, and of a fine light-yellow colour.

33. *Citrus australasica*, F.M. (Native Lime.)—Diameter, 6 to 10 inches ; height, 15 to 20 feet. A low-sized and beautiful tree, growing in the scrubs on the Darlington Range. The trunk is erect and well diffused, with small branches bearing fruit about two inches long, and of an oblong form. The wood is close-grained, hard, and of a yellow colour.

34. *Atlantia glauca*, Hook. (The Native Kumquat)—Diameter, 2 to 6 inches ; height, 8 to 15 feet. A small tree or shrub, armed with straight or incurved axillary spines of a quarter of an inch long upon the branches. The fruit is globular, about half an inch in diameter, and produces an agreeable beverage from its acid juice. The wood is close grained, and takes a fine polish. Found in great abundance in the Darling Downs and the Maranoa districts.

35. *Glycosmis pentaphylla*, Corr.—Diameter, 4 to 6 inches ; height, 8 to 14 feet. A small tree occurring sparingly in the scrubs at the entrance of the Mary River. Wood close-grained.

36. *Acronychia Baueri*, Schott.—Diameter, 6 to 12 inches ; height, 16 to 24 feet. Small-sized tree, found in the scrubs bordering the coast. Wood close-grained, but not used.

37. *Acronychia imperforata*, F.M.—Diameter, 12 to 15 inches ; height, 20 to 30 feet. A handsome shady tree, occurring in the scrubs on the coast. Timber fine-grained, easily wrought, but not much used.

38. *Acronychia lævis*, Frost.—Diameter, 15 to 18 inches ; height, 30 to 40 feet. A tall slender tree; timber not used.

39. *Pentaceras australis*, Hook, Scrub White Cedar of the colonists.—Diameter, 12 to 24 inches ; height, 40 to 60 feet. Occurs principally in the scrubs near the coast. The timber is close-grained, tough, and firm.

40. *Zanthoxylum brachyacanthum*, F.M. (Satinwood).—Diameter, 6 to 12 inches; height, 20 to 30 feet. Found in small quantities in most of the coast scrubs of Queensland. The timber is close-grained, of a yellow colour, and susceptible of a high polish.

41. *Geijera parviflora*, Lindl.—Diameter, 6 to 12 inches ; height, 20 to 30 feet. Tree occurring in many of the brigalow scrubs. Timber hard and close-grained ; it is, however, apt to split in seasoning.

42. *Geijera Muelleri*, Benth. (Balsam Capivi Tree).—Diameter, 12 to 13 inches; height, 40 to 60 feet. This handsome and glabrous tree is dispersed through the brigalow and Araucaria scrubs in the East and West Moreton districts. The timber is nicely marked, and of agreeable fragrance when green.

43. *Geijera salicifolia*, Schott.—Diameter, 10 to 15 inches; height, 30 to 40 feet. The long slender drooping branches with their long narrow leaves, have a very beautiful appearance. It occurs in brigalow scrubs on the coast, and in the interior; wood is hard, close-grained, and nicely marked.

44. *Evodia micrococca*, F.M.—Diameter, 10 to 15 inches; height, 20 to 30 feet. A small-sized tree, of no great beauty; met with in the scrubs on the banks of the rivers in the Moreton district.

CELASTRINEÆ.

45. *Celastrus dispermus*, F.M.—Diameter, 3 to 5 inches; height, 12 to 16 feet. A small-sized glabrous tree of some beauty when not over-crowded with other trees; wood close-grained, and takes a fine polish.

46. *Siphonodon australe*, Benth.—Diameter, 10 to 24 feet; height, 40 to 50 feet. A handsome scrub tree of frequent occurrence; wood close-grained, of a yellowish colour.

47. *Denhamia pittosporoides*, F.M.—Diameter, 6 to 8 inches; height, 20 to 30 feet. Slender tree, found on the borders of scrubs inland; timber is hard, fine grained, and takes a good polish.

48. *Denhamia obscura*, Meisn.—Diameter, 3 to 4 inches; height, 12 to 15 feet. Small tree, found in brigalow scrubs near Ipswich; wood fine grained and tough.

49. *Elaeodendron australe*, Vent.—Diameter, 4 to 12 inches; height, 20 to 30 feet. Slender-growing, glabrous tree; timber close-grained and prettily marked.

RHAMNEÆ.

50. *Alphitonia excelsa*, Reissek. (Mountain or Red Ash).—Diameter, 18 to 24 inches; height, 45 to 60 feet. This valuable tree is plentiful in the forest and in the scrubs, both on the coast and in the interior. The timber is hard, close-grained, and durable; takes a high polish, and is suitable for gun-stocks and a variety of other purposes.

PITTOSPOREÆ.

51. *Pittosporum rhombifolium*, A.Cunn.—Diameter, 6 to 12 inches; height, 40 to 55 feet. A fine tree with glossy foliage, found in the scrubs on the Brisbane Rriver. The wood is of a white colour; not used.

52. *Hymenosporum flavum* F.M.—Diameter, 6 to 12 inches; height, 20 to 46 feet. Found in scrubs in East and West Moreton. Wood close-grained, of a white colour.

53. *Pittosporum phillyraeoides*, D.C.—Diameter, 4 to 6 inches; height, 20 to 25 feet. Met with in the brigalow scrubs; wood close-grained, and of a white colour.

CAPPARIDEÆ.

54. *Capparis nobilis.* F.M. (Native Pomegranate).—Diameter, 6 to 14 inches; height, 20 to 25 feet. A small tree. The timber is hard and close-grained.

55. *Cappar is Mitchelli,* Lindl. (Small Native Pomegranate).— Diameter, 6 to 12 inches; height, 14 to 30 feet. This much-branched small tree is very plentiful in the brigalow scrubs in the Darling Downs; wood very hard.

56. *Apophyllum anomalum,* F.M.—Diameter, 6 to 16 inches; height, 20 to 30 feet. A shrub or small tree, almost leafless, with cylindrical, often pendulous branches; occurring in the brigalow scrubs in the Darling Downs District. Wood very hard.

STERCULIACEÆ.

57. *Sterculia quadrifida,* R. Br.—Diameter, 18 to 24 inches; height, 30 to 40 feet. A moderate sized deciduous tree; wood soft and spongy; the bark is used for nets and fishing lines.

58. *Tarrietia argyrodendron,* Benth. (Silver Tree.)—Diameter, 18 to 30 inches; height, 40 to 60 feet. A large tree growing in great quantities in the scrubs bordering the banks of the River Brisbane; the timber is not much used.

59. *Tarrietia,* Sp., F. M.—Diameter, 18 to 24 inches; height, 60 to 70 feet. Plentiful in the coast scrubs; timber is tough and close-grained, but seldom used.

60. *Commersonia echinata.* Frost.—Diameter, 6 to 12 inches; height, 20 to 30 feet. A tall shrub or small tree, of general occurrence on the banks of rivers; the aboriginals use the fibre of the bark for kangaroo and fishing nets; timber not used.

LINEÆ.

61. *Erythroxylon australe,* F. M.—Diameter, 6 to 12 inches; height, 20 to 30 feet. This shrub or small-sized tree occurs in considerable abundance in the brigalow scrub near Ipswich; wood hard, fine-grained, it takes a good polish, can be used for cabinet work.

SAPINDACEÆ.

62. *Harpullia pendula,* Planch. (Tulip-wood).—Diameter, 14 to 24 inches; height, 50 to 60 feet. Found in some abundance on the alluvial banks of rivers. The timber is close-grained, firm, and beautifully marked, and is much esteemed for cabinet work.

63. *Cupania xylocarpa,* A. Cunn.—Diameter, 12 to 24 inches; height, 40 to 50 feet. Moderate tree in good situations. Timber close-grained and hard, particularly so when dry.

64. *Cupania serrata,* F. M.—Diameter, 8 to 14 inches; height, 20 to 30 feet. Plentiful in the scrubs on the banks of rivers. Timber close-grained.

65. *Cupania anacardioides.* A. Rich.—Diameter, 12 to 18 inches; height, 30 to 40 feet. Slender tree in considerable abundance on the alluvial banks of rivers. Timber seldom used.

66. *Cupania pseudorhus,* A. Rich.—Diameter, 14 to 20 inches; height, 30 to 50 feet. Tree of moderate size; growing in great

abundance in the scrubs bordering the coast. Timber fine-grained and beautiful.

67. *Oupania semiglauca*, F.M.—Diameter, 10 to 18 inches; height, 30 to 60 feet. Middle-sized tree, wood soft, and as yet of no recognised value.

68. *Diploglottis Cunninghamii*, Hook. (Native Tamarind).—Diameter, 12 to 24 inches; height, 40 to 55 feet. Timber seldom used, though compact and durable.

69. *Ratonia pyriformis*, Benth.—Diameter, 10 to 18 inches; height, 30 to 45 feet. Moderate-sized tree; timber firm and close-grained.

70. *Nephelium tomentosum*. F.M.—Diameter, 10 to 15 inches; height, 30 to 40 feet. Small sized tree; timber not used.

71. *Heterodendron diversifolium*, F.M.—Diameter, 4 to 6 inches; height, 10 to 15 feet. Common in brigalow scrubs; wood of a reddish colour. Its great strength renders it fit for pick handles.

72. *Dodonæa triquetra*, Andr. (Hop Bush).—Diameter, 3 to 4 inches; height, 10 to 12 feet. Branching shrub; wood hard, close-grained.

73. *Atalaya salicifolia*, Blume.—Diameter, 14 to 22 inches; height, 30 to 50 feet. A handsome glabrous, green or somewhat glaucous tree, occurring in the alluvial scrubs on the coast. Timber close grained, hard, and takes a good polish.

ANACARDIACEÆ.

74. *Rhus rhodanthema*, F.M. (Dark Yellow-wood).—Diameter, 18 to 24 inches; height, 50 to 70 feet. A picturesque tree, of general occurrence in the scrubs on the banks of rivers; the trunk is of moderate size, covered with a rough scaly bark; the branches are small and numerous, the leaves are pinnate, the flowers are red. The wood is soft, fine-grained, and beautifully marked, and is much esteemed for cabinet work. At the present time the sawyers are receiving 10s. to 12s. per hundred feet.

75. *Spondias pleiogyna*, F.M.—Diameter, 24 to 36 inches; height, 40 to 60 feet. A moderate-sized tree of rare occurrence in the coast scrubs. Wood soft when cut, but afterwards becomes hard and tough; not as yet used.

ARALIACEÆ.

76. *Panax elegans*, F.M.—Diameter, 12 to 16 inches; height, 30 to 40 feet. A singular, moderate-sized tree, with magnificent, large, simple, or doubly pinnate leaves, occurring in dense scrubs. Wood light, soft, and of little durability.

RUBIACEÆ.

77. *Sarcocephalus cordatus*, Miq. (Leichhardt's Tree).—Diameter, 24 to 30 inches; height, 40 to 60 feet. A handsome and beautiful moderate-sized tree, found on the alluvial banks of the Don River, Port Denison, &c., &c. Its wood is soft, close-grained, and takes a good polish; often used for building and other purposes.

78. *Ixora Pavetta*, Roxb.—Diameter, 2 to 4 inches; height, 8 to 10 feet. A beautiful flowering shrub found in the borders of scrubs; wood very hard and fine-grained.

79. *Hodgkinsonia ovatiflora,* F. M.—Diameter, 5 to 10 inches : height, 12 to 20 feet. Small slender tree; wood close-grained.

80. *Canthium lucidum,* Hook and Arn.—Diameter, 6 to 12 inches; height, 20 to 30 feet. Small tree; wood hard and close-grained, and takes a good polish.

81. *Canthium oleifolium,* Hook.—Diameter, 5 to 10 inches; height, 25 to 30 feet. In brigalow scrubs, near Ipswich; wood hard, close-grained, and capable of a high polish.

82. *Canthium vacciniifolium,* F. M.—Diameter, 1 to 4 inches; height, 6 to 10 feet: wood, close-grained, used for walking-sticks.

MYRTACEÆ.

83. *Eucalyptus pilularis,* Sm. (Blackbutt).—Diameter, 48 to 96 inches: height, 60 to 100 feet. A large tree, with a dark-coloured rough and somewhat furrowed persistent bark, at least on the trunk and on the main branches, that of the smaller branches smooth and deciduous; furnishes excellent timber for house carpentry, or any purposes where strength and durability are required. Market value, 7s. to 8s. per hundred feet.

84. *Eucalyptus hæmustoma,* Sm. (Spotted Gum).—Diameter, 24 to 28 inches; height, 60 to 120 feet. A very large tree, with a smooth deciduous bark, leaving a spotted or variegated trunk; considered a first-class timber for shipbuilding, and much used for wheelwrights' work and other purposes. Market value, 7s. to 8s. per hundred feet.

85. *Eucalyptus microcorys,* F. M. (Peppermint Tree).—Diameter, 18 to 36 inches; height, 60 to 80 feet. A tall tree, with a persistent furrowed fibrous bark, occurring in the forest near the Logan, Brisbane, and Pine Rivers; timber strong and durable, used by wheelwrights for naves, felloes, and spokes.

86. *Eucalyptus hemiphloia,* F. M. (Yellow Box).—Diameter, 20 to 40 inches; height, 50 to 60 feet. A moderate-sized tree, producing an excellent timber, famous for its hardness, toughness, and durability, market value, from 7s. to 8s. per hundred feet.

87. *Eucalyptus siderophloia,* Benth. (Ironbark).—Diameter, 20 to 40 inches; height, 70 to 100 feet. A tall tree, with a hard, persistent, rough and furred bark. Occupant of many ridgy, stony, forest lands in East and West Moreton and Darling Downs districts. This timber has the highest reputation for strength and durability, and is used for large beams in building stores for heavy goods, railway sleepers, and other purposes where great strength is required. Market value, 7s. to 8s. per hundred feet.

88. *Eucalyptus maculata,* Hook. (Spotted Gum).—Diameter, 20 to 36 inches; height, 60 to 80 feet. A lofty tree, with a smooth bark falling off in patches so as to give the trunk a spotted appearance. It is a very valuable timber, highly prized for many purposes on account of its strength and elasticity, used for buggy shafts, cogs of wheels, &c. In bridge-building it is used for members under tension, and is found to have the highest constant strength of any of the Queensland timbers. Market value, 7s. to 8s. per hundred feet.

89. *Eucalyptus saligna,* Sm. (Grey Gum).—Diameter, 24 to 40 inches; height, 60 to 90 feet. A tall tree with a smooth silver-grey shining bark, shedding in thick longitudinal strips. Of frequent occurrence on the forest ridges near the Brisbane River. Useful

timber ; in good repute for rails in fencing, as it does not readily take fire, and building purposes, being both strong and durable. Market value, from 7s. to 8s. per hundred feet.

90. *Eucalyptus resinifera,* Sm. (Red Mahogany).—Diameter, 20 to 30 inches ; height, 60 to 80 feet. A tall tree, with a rough persistent bark on the trunk, but more or less deciduous on the branches. The timber is much prized for its strength and durability, and is used for piles, as it is said to resist the action of cobra. Market value, from 7s. to 8s. per hundred feet.

91. *Eucalyptus corymbosa,* Sm. (Bloodwood).—Diameter, 24 to 30 inches ; height, 50 to 60 feet. A fair-sized tree, with a persistent furred bark ; timber subject to gum veins, but very durable, principally used for posts, does not readily take fire or suffer much from white ants. Market value, from 7s. per hundred feet.

92. *Eucalyptus botryoides,* Sm. (Blue Gum).—Diameter, 30 to 50 inches ; height, 70 to 100 feet. A lofty spreading tree, with a rough furrowed persistent bark. Of frequent occurrence both upon the coast and in the interior ; a valuable timber, hard, tough, and durable. The only timber used for felloes of wheels, and one of the finest timbers for ship-building. Market value, 8s. to 9s. per hundred feet.

93. *Eucalyptus tereticornis,* Sm. (Red Gum).—Diameter, 18 to 36 inches ; height, 60 to 90 feet. A fair-sized tree ; timber used in fencing, building, plough-beams, poles and shafts of drays, and also in ship-building. Market value, from 7s. to 8s. per hundred feet

94. *Eucalyptus rostrata,* Schlecht. (Flooded Gum).—Diameter, 40 to 48 inches ; height, 80 to 100 feet. A majestic tree, inhabiting the rich alluvial flats upon the banks of the rivers, and in such localities has a pillar-like trunk clear of branches for three-fourths of its entire height. The timber is of high repute for strength and durability. The market value at the present time is from 8s. to 9s.

95. *Eucalyptus fibrosa,* F.M. (Stringybark).—Diameter, 18 to 36 inches ; height, 40 to 75 feet. Timber much prized for flooring boards, and of considerable strength and durability. Market value, from 6s. to 7s. per hundred feet.

96. *Eucalyptus tessellaris,* F.M. (Moreton Bay Ash).—Diameter, 14 to 24 inches ; height, 30 to 60 feet. Timber of brownish colour, not hard but tough ; highly spoken of for building purposes in the northern parts of the colony. Market value, from 7s. to 8s. per hundred feet.

97. *Eucalyptus melanophloia,* F.M. (Silver-leaved Ironbark).—Diameter, 18 to 24 inches ; height, 30 to 70 feet. Middle-sized tree, with a blackish, persistent, deeply furrowed bark, the foliage more or less glaucous or mealy white : timber used for railway sleepers, and for fencing and other purposes. Market value, from 8s. to 10s. per hundred feet.

98. *Eucalyptus crebra,* F.M. (White narrow-leaved Ironbark).—Diameter, 20 to 36 inches ; height, 70 to 90 feet. A fair-sized tree, with a hard, blackish, rough persistent bark ; producing an excellent timber—hard, tough, and durable.—valuable for building purposes. Market value, from 8s. to 9s. per hundred feet.

99. *Eucalyptus leptophleba,* F.M.—Diameter, 18 to 36 inches ; height, 50 to 80 feet. Moderate-sized or large tree ; timber hard and durable.

100. *Eucalyptus citriodora*, Hook. (Citron-scented Gum).—
Diameter, 18 to 34 inches; height, 40 to 70 feet. A middle-sized tree
with a smooth bark. The foliage emitting a strong odour of citron
when rubbed. Timber hard and durable; used for house carpentry.

101. *Eucalyptus platyphylla*, F.M.—Diameter, 12 to 18 inches;
height, 40 to 50 feet. A handsome tree, with light-green foliage and
smooth white deciduous bark. The timber is dark, close-grained;
excellent to put in damp places.

102. *Eucalyptus acmenioides*, Schau.—Diameter, 18 to 30 inches;
height, 40 to 60 feet. Of frequent occurrence in the forests on the
coast. Timber heavy, strong, and durable; has been found good for
flooring-boards and other purposes.

103. *Eucalyptus Planchoniana*, F.M.—Diameter, 18 to 30 inches:
height, 50 to 80 feet. A tall tree, producing an excellent timber;
valuable for wheelwright and other purposes.

104. *Eucalyptus Baiieyana*, F.M.—Diameter, 20 to 30 inches:
height, 40 to 70 feet. A moderate-sized tree, with a furrowed and
fibrous Vandyke brownish-coloured bark. Of the quality of the
timber hardly anything is known.

105. *Callistemon lanceolatus*, D.C. (Bottle Brush Tree).—
Diameter, 12 to 18 inches; height, 30 to 40 feet. Small tree, growing in or near the beds of rivers; wood hard and heavy; is used for
ship-building, wheelwrights' and many implements, such as mal
lets, &c.

106. *Callistemon salignus*, D.C. (Broad-leaved Tea Tree).—
Diameter, 18 to 24 inches; height, 40 to 60 feet. Wood, hard and
close-grained; very durable undergound.

107. *Melaleuca genistifolia*, Sm.—Diameter, 20 to 24 inches;
height, 30 to 40 feet. Moderate-sized tree, wood close-grained, hard,
and durable.

108. *Melaleuca leucadendron*, Linn. (White Tea Tree).—Diameter, 24 to 40 inches; height, 40 to 80 feet. Moderate-sized tree,
timber hard and close-grained, excellent for posts in damp ground and
piles for wharves, and is said to be imperishable underground.

109. *Melaleuca styphelioides*, Sm. (Prickly-leaved Tea Tree).—
Diameter, 4 to 10 inches; height, 20 to 30 feet. Small-sized tree,
timber hard, close-grained; stands well in damp situations. It has
been said that this timber has never been known to decay.

110. *Melaleuca linariifolia*, Sm.—Diameter, 10 to 14 inches;
height, 30 to 40 feet. Small-sized tree; timber hard, close, and
durable.

111. *Melaleuca nodosa*, Sm. (Tea Tree).—Diameter, 10 to 20 inches
height, 30 to 40 feet. Small tree, qualities same as 102 and 103.

112. *Tristania conferta*, R. Br. (Box).—Diameter, 35 to 50
inches; height, 80 to 100 feet. A large spreading tree, with a smooth
brown deciduous bark, and dense foliage; very generally distributed
on open forest ground. The timber is much prized for its strength and
durable qualities. Market value, from 8s. to 9s. per hundred feet.
Used in ship-building; ribs of vessels from this tree have lasted unim-
paired thirty years and more.

113. *Tristania suaveolens*, Sm.—A small but handsome tree, found
along the banks of fresh-water streams; timber very close and elastic;
used for carpenters' mallets and cogs of wheels in machinery.

114. *Syncarpia laurifolia*, Ten. (Turpentine or Pearbbie of Frazer's Island).—Diameter, 48 to 96 inches; height, 70 to 100 feet. A magnificent tree, with a rough persistent bark on the trunk, but more or less deciduous on the branches. Timber valuable for piles and posts for timber fences; very durable underground, and said to resist the *Teredo navalis* in salt water. Market value, 7s. to 8s. per hundred feet.

115. *Lysicarpus ternifolius*, F.M.—Diameter, 18 to 24 inches; height, 40 to 50 feet. A middle-sized tree, with a soft thick fibrous bark. Timber is hard, heavy, and elastic, prettily marked; used for cabinet work, but more particularly for piles, bridges, railway sleepers, &c.

116. *Backhousia myrtifolia*, Hook and Harv.—Diameter, 12 to 18 inches; height, 20 to 40 feet; small tree; timber close-grained and prettily marked.

117. *Backhousia citriodora*, F.M.—Diameter, 6 to 12 inches; height, 20 to 30 feet. Small-sized tree, wood hard and fine-grained; useful for several purposes.

118. *Rhodomyrtus psidioides*, Benth.—Diameter, 12 to 20 inches; height, 30 to 40 feet. Frequent in the scrubs; wood close-grained; not much known.

119. *Myrtus Hillii*, Benth. (Scrub Ironwood).—Diameter, 6 to 12 inches; height, 20 to 40 feet. Small-sized tree; wood remarkably hard.

120. *Myrtus Bidwillii*, Benth.—Diameter, 6 to 10 inches; height, 15 to 20 feet. Timber, close-grained; not much known.

121. *Myrtus acmenioides*, F.M.—Diameter, 12 to 18 inches; height, 30 to 40 feet. Small tree, frequent in the scrubs; timber, close-grained; not much used.

122. *Rhodamnia trinervia*, Blume.—Diameter, 10 to 18 inches; height, 20 to 30 feet. Small tree; wood, close-grained and firm.

123. *Rhodamnia argentea*, Benth.—Diameter, 15 to 22 inches; height, 40 to 60 feet. Found in great abundance in moist low scrubs; wood, tough and firm.

124. *Eugenia Ventenatii*, Benth.—Diameter, 18 to 20 inches; height, 40 to 60 feet. Of frequent occurrence in moist scrubs; wood close-grained and of a pinkish hue.

125. *Eugenia Smithii*, Poir. (Lilly Pillies).—Diameter, 12 to 18 inches; height, 30 to 40 feet; wood close, but apt to split in seasoning.

126. *Eugenia angophoroides*, F.M.—Diameter, 12 to 18 inches; height, 30 to 40 feet. Small-sized tree; timber but little known.

127. *Eugenia grandis*, Wight.—Diameter, 20 to 30 inches; height, 40 to 70 feet. A large and handsome tree; timber close-grained; not much known.

128. *Leptospermum flavescens*," Sm.—Diameter, 8 to 5 inches; height, 15 to 20 feet. Frequent about fresh-water creeks; wood hard and close-grained.

PROTEACEÆ.

129. *Grevillea robusta*, A. Cunn. (Silky Oak).—Diameter, 30 to 40 inches; height, 80 to 100 feet. A lofty tree, of frequent occurrence in the scrubs along the coast, and for a considerable distance in the interior. The wood is extensively used for staves for tallow casks and

is in much repute for cabinet work. At the present the sawyers are receiving at the rate of 8s. to 9s. per hundred feet.

130. *Grevillea Hilliana*, F.M.—Diameter, 10 to 18 inches; height, 40 to 60 feet. A beautiful tree, found in dense scrubs on the banks of the Logan and Albert Rivers. Wood easily wrought; not used.

131. *Grevillea gibbosa*, R. Br.—Diameter, 8 to 12 inches; height, 20 to 30 feet. A small tree, occurring in the forest ground on the bank of the Don River. Wood fine-grained and nicely marked.

132. *Grevillea polystachya*, R. Br.—Diameter, 12 to 18 inches; height, 30 to 45 feet. A handsome tree; timber strong, durable, prettily marked; used for cabinet work, &c.

133. *Macadamia ternifolia*, F.M. (Queensland Nut).—Diameter, 3 to 12 inches; height, 30 to 50 feet. A small-sized tree, with a very dense foliage. Found in dense, moist scrubs on the banks of rivers; wood firm, fine-grained, and takes a good polish. This tree bears an edible nut of excellent flavor, which is relished by the white colonists as well as by the aborigines. It forms a nutritious article of food to the latter, and, in consequence, the same restriction with regard to this tree as in the case of *Araucaria Bidwillii* (Bunya Bunya), is made in the licenses issued for cutting timber.

134. *Stenocarpus sinuatus*, Endl. (Tulip Tree).—Diameter, 18 to 30 inches; height, 40 to 50 feet. A charming moderate-sized tree, with dense, bright, glossy foliage. It occurs often in scrubs, some distance from the coast. The wood is nicely marked, and would admit of a good polish.

135. *Stenocarpus salignus*, R. Br. (Beefwood).—Diameter, 18 to 24 inches; height, 30 to 50 feet. A middle-sized beautiful tree. Timber is red-coloured, close in the grain, hard, and splits easily; valuable for the finer kinds of coopers' work.

136. *Orites excelsa*, R. Br.—Diameter, 6 to 14 inches; height, 30 to 40 feet. Of frequent occurrence in the scrubs bordering the coast. Timber hard, nicely marked, and takes a good polish.

137. *Banksia integrifolia*, Linn. (Beefwood).—Diameter, 8 to 12 inches; height, 20 to 30 feet. Occurring on sandy ridges near the coast, and for a considerable distance inland.

138. *Banksia æmula*, R. Br.—Diameter, 9 to 15 inches; height, 20 to 30 feet. Small-sized tree, found on sandy ridges on the coast; wood prettily marked.

139. *Xylomelum pyriforme*, Knight. (Native Pear).—Diameter, 3 to 6 inches; height, 12 to 18 feet. Small-sized tree, occurring on sandy ridges; wood nicely marked.

RHIZOPHOREÆ.

140. *Bruguiera Rheedii*, Blume. (Mangrove).—Diameter, 6 to 12 inches; height, 12 to 20 feet. Small tree; timber handsome; bark is astringent and used for tanning purposes.

SANTALACEÆ.

141. *Exocarpus latifolia*, Br. (Broad-leaved Cherry Tree).—Diameter, 6 to 9 inches; height, 15 to 25 feet. Small tree, of frequent occurrence in scrubs on the coast; the timber is very hard and fragrant; excellent for cabinet work.

142. *Exocarpus cupressiformis*. Labill. (Cherry Tree).—Diameter, 4 to 8 inches; height, 10 to 16 feet. Small tree, found sparingly on open forest ground. The wood is close-grained and handsome.

143. *Santalum lanceolatum*, R. Br. (Sandalwood).—Diameter, 3 to 6 inches: height, 15 to 25 feet. A small tree. sparingly distributed through the brigalow scrubs. The wood is close-grained, and takes a good polish.

MYOPORINEÆ.

144. *Eremophila Mitchelli*, Benth. (Bastard Sandalwood).—Diameter, 6 to 12 inches; height, 20 to 30 feet. Small tree, of frequent occurrence in open forest land in the Darling Downs district. Timber very hard, beautifully grained and very fragrant; makes handsome veneers for cabinet-work.

145. *Eremophila bignoniæfolia*, F.M.—Diameter, 6 to 9 inches; height, 20 to 30 feet. A small tree, sparingly distributed through the brigalow scrubs. The wood is white, close-grained, and takes a good polish.

146. *Myoporum platycarpum*, R. Br.—Diameter, 12 to 15 inches; height, 15 to 25 feet. A small tree, wood close-grained, not used.

VERBENACEÆ.

147. *Avicennia officinalis*, Linn. (Mangrove).—Diameter, 12 to 24 inches; height, 20 to 30 feet. Found on salt-water estuaries; timber used for knees of boats, stonemasons' mallets, etc.

148. *Gmelina Leichhardtii*, F.M. (Beech) —Diameter, 24 to 42 inches; height, 80 to 120 feet. Found in the scrubs bordering the rivers on the coast. A very useful timber, strong and durable, and easily worked: it does not expand by damp or contract by dry weather; much prized for the decks of vessels and the flooring of verandahs.

149. *Vitex lignum-vitæ*, A. Cunn. (Lignum Vitæ).—Diameter, 20 to 24 inches; height, 50 to 70 feet. A tree of general occurrence in the moist low scrubs bordering the coast. The timber is hard, close-grained, and of a blackish colour, used by cabinet-makers, &c.

MONIMIACEÆ.

150. *Daphnandra micrantha*, Benth.—Diameter, 18 to 30 inches; height, 60 feet to 80 feet. Moderate-sized tree, occasionally found in low moist scrubs. Timber quite yellow when fresh, takes a fine polish, and is easily worked.

LAURINEÆ.

151. *Endiandra pubens*, Meissn.—Diameter, 19 to 24 inches; height, 40 to 70 feet. Moderate-sized tree, of general occurrence in the scrubs on the banks of the Brisbane and Albert Rivers. Timber not as yet used.

151A. *Endiandra glauca*, R. Br.—Diameter, 12 to 24 inches; height, 30 to 70 feet. Of frequent occurrence in coast scrubs. Wood fragrant, soft, and easily worked.

152. *Tetranthera ferruginea*. R. Br.—Diameter, 14 to 20 inches; height, 30 to 40 feet. Wood close-grained, not used.

153. *Cryptocarya patentinercis*, F. M.—Diameter, 12 to 20 inches height, 50 to 40 feet. A small-sized tree; timber of apparent value, but not used for any purpose.

153A. *Cryptocarya glauce scens*, R. Br. (White Laurel).—Diameter, 12 to 20 inches; height, 40 to 60 feet. The wood has an aromatic fragrance; easily worked, and said to be durable.

TILIACEÆ.

154. *Elæocarpus obovatus*, G. Don.—Diameter, 12 to 20 inches; height, 30 to 40 feet. A tree common in the scrubs on the banks of the Brisbane River; wood fine-grained, not yet used.

155. *Elæocarpus grandis*, F. M. (Calham).—Diameter, 24 to 36 inches; height, 80 to 90 feet. This tree is frequent in the moist low scrubs along the coast, the trunk is erect, the bark smooth, the branches, with their thin, bright-green, glossy foliage, are thinly scattered over its lofty head: the wood is soft and easily worked, it is likely to be serviceable for breaks for railway carriages.

156. *Grewia latifolia*, F. M.—Diameter, 6 to 8 inches; height, 10 to 20 feet. Wood hard, close-grained, and takes a good polish.

CORNACEÆ.

157. *Marlea vitiensis*, Benth. (Musk tree).—Diameter 6 to 12 inches; height, 20 to 30 feet. A small-sized tree, generally with a gouty trunk; bark dark-coloured, rough and scaly; wood bright-yellow, with a fine undulating appearance, black at the centre: found in moist low scrubs.

JASMINEÆ.

158. *Olea paniculata*, R. Br. (Native Olive).—Diameter, 18 to 24 inches; height, 50 to 70 feet. A moderate-sized tree, of frequent occurrence in the scrubs both on the coast and also in the interior; timber close-grained, hard, and durable.

159. *Notelæa ovata*, R. Br. (Dunga Vunga).—Diameter, 6 to 12 inches; height, 20 to 30 feet. A slender tree, found in scrubs; wood close-grained.

160. *Notelæa longifolia*, Vent.—Diameter, 12 to 18 inches; height, 30 to 40 feet. A small tree; timber hard, closed-grained.

161. *Notelæa microcarpa*, R. Br.—Diameter, 9 to 12 inches; height, 30 to 45 feet. A tree of frequent occurrence on the border of scrubs on the coast; wood hard and close-grained.

EUPHORBIACEÆ.

162. *Mallotus claoxyloides*, M. Ar.—Diameter, 9 to 16 inches; height, 15 to 30 feet. Occurring both in moist, low scrubs and in dry rocky places. Timber white, hard, and close-grained.

163. *Mallotus philippinensis*, M. Ar.—Diameter, 6 to 14 inches; height, 30 to 45 feet. Small tree, generally found in rich scrubs. Wood close-grained and very tough.

164. *Mallotus nesophilus*, F.M.—Diameter, 12 to 18 inches; height, 35 to 45 feet. Of frequent occurrence in low and moist scrubs on the coast. Wood of a uniform white colour, soft and easily worked.

165. *Petalostigma quadriloculare*, F. M. (Crab Tree).—Diameter, 12 to 18 inches; height, 40 to 50 feet. Found in great abundance growing on poor sandy soil in the open forest. The timber is hard and fine-grained, and promises to be useful to the cabinet-makers. The bark possesses a powerful bitter and qualities somewhat similar to Peruvian bark.

166. *Excæcaria agallocha*, Linn. (River Poisonous Tree).—Diameter, 6 to 18 inches; height, 20 to 30 feet. Found on the estuaries of saltwater creeks and rivers. Produces by incision in the bark an acrid, milky juice, which is so volatile that nobody, however careful, can gather a quarter of a pint without being affected. The symptoms are an acrid burning sensation in the throat, sore eyes and head-ache; a single drop falling into the eyes it is believed will cause loss of sight. The natives of Eastern Australia, as well as those of New Guinea, Fiji, &c., use this poisonous juice to cure certain ulcerous chronic diseases (Morrell's testimony). Wood light, white, and soft; will answer for carving and marqueterie.

166A. *Excæcaria Dallachyana*, Baill. (Scrub Poison Tree).—Diameter, 12 to 14 inches; height, 40 to 50 feet. A slender, spreading tree often met with in brigalow scrubs; the wood is soft, fine-grained and elastic. The juice is pure white, and nauseous in taste; a single drop falling into the eye will injure the sight. If properly prepared, a gum-elastic could be made from it.

167. *Briedelia exaltata*, F. M.—Diameter, 12 to 18 inches; height, 33 to 45 feet. Not unfrequent in moist, low scrubs on the coast. Timber hard and close-grained.

168. *Dissiliaria baloghioides*, F. M. (Teak).—Diameter, 18 to 30 inches; height, 40 to 60 feet. Moderate-sized tree, found in great abundance in the coast scrubs. Timber hard, close-grained, and durable.

169. *Mallotus discolor*, F.M.—Diameter, 12 to 18 inches; height, 35 to 45 feet. Of frequent occurrence in low moist scrubs along the coast. Wood of a uniform white colour, soft, and easily worked.

170. *Croton Verreauxii*, Baill. (Cascarilla).—Diameter, 8 to 12 inches; height, 30 to 40 feet. A small-sized tree, bark grey and rough, with a red sap. Wood of a yellowish-white colour; soft, and of no value; the bark contains an agreeable bitter.

171. *Baloghia lucida*, Endl. (Scrub Bloodwood).—Diameter, 8 to 16 inches; height, 30 to 40 feet. A small-sized tree, abundant in the coast scrubs; timber not used.

URTICEÆ.

172. *Celtis ingens*, F.M.—Diameter, 6 to 12 inches; height, 25 to 35 feet. A small tree of frequent occurrence in the coast scrubs. Wood white, soft, and pliable; used for hoops for casks.

173. *Aphananthe philippinensis*, Plan.—Diameter, 4 to 12 inches; height, 20 to 40 feet. This species abounds in moist scrubs. Timber not used.

174. *Laportea photiniphylla*, Wedd. (Nettle Tree).—Diameter, 15 to 24 inches; height, 30 to 50 feet. A beautiful tree; wood soft, spongy; the bark is used for making fishing nets.

175. *Ficus asper*, Forst. (Rough-leaved Fig).—Diameter, 9 to 18 inches; height, 30 to 45 feet. Timber of no apparent value.

176. *Ficus vesca*, F.M. (Clustered Fig Tree).—Diameter, 18 to 48 inches; height, 50 to 80 feet. Found on the alluvial banks of rivers and creeks. The fruit, which is of a light red colour when ripe, hangs in clusters along the trunk and on some of the largest branches, and is used by the aborigines.

177. *Ficus macrophylla*, Desf. (Moreton Bay Fig.)—Diameter, 36 to 76 inches; height, 50 to 100. A large and magnificent wide-spreading tree; yielding its milk sap copiously for caoutchouc.

178. *Cudrania Javanensis*, Trecul. (Cockspur Thorn)—A rambling, thorny climber; Duramen or heartwood, dark yellow colour, hard, and used in dyeing yellow and brown.

179. *Pseudo-morus Brunoniana*, Bureau.—Diameter, 6 to 15 inches; height, 15 to 20 feet. A small, handsome spreading tree, with a milky juice ; found in scrubs on the banks of the Brisbane River, &c.

SAPOTACEÆ.

180. *Hormogyne cotinifolia*, D.C.—Diameter, 6 to 9 inches; height, 20 to 30 feet. Small tree, wood close-grained.

181. *Chrysophyllum pruniferum*, F.M.—Diameter, 12 or 20 inches ; height, 30 to 70 feet. Moderate-sized tree, sparingly distributed over moist, low scrubs. Wood of a uniform pale yellow colour, close-grained.

182. *Achras Pohlmanniana*, F.M.—Diameter, 8 to 18 inches; height, 20 to 60 feet. Small tree, frequent in low moist scrubs. Timber hard and close-grained.

183. *Achras myrsinoides*, A. Cunn.—Diameter, 8 to 12 inches; height, 20 to 35 feet. Small tree, frequent in low moist scrubs. Timber hard and close-grained.

184. *Achras obovata*, F.M.—Diameter, 8 to 12 inches; height, 20 to 35 feet. Small tree, frequent in low moist scrubs. Timber hard and close-grained.

EBENACEÆ.

185. *Cargillia australis*, R. Br.—Diameter, 6 to 12 inches; height, 30 to 40 feet. Timber very tough and firm, and likely to be used for many purposes.

186. *Maba fasciculosa*, F.M.—Diameter, 18 to 24 inches; height, 60 to 80 feet. Of common occurrence in the scrubs bordering on river banks. Wood tolerably close-grained.

187. *Maba geminata*, R. Br.—Diameter 9 to 12 inches ; height, 50 to 60 feet. Slender tree, found growing in scrubs. Wood soft and tough.

188. *Diospyros hebecarpa*, A. Cunn.—Diameter, 12 to 18 inches ; height, 30 to 50 feet. Handsome tree, occurring in moist scrubs on the coast. Timber soft, elastic ; used for pick handles, &c., &c.

SAXIFRAGEÆ.

189. *Ceratopetalum apetalum*, Don. (Coach-wood.)—Diameter, 24 to 36 inches, height, 70 to 90 feet. A beautiful tree, with long cylindrical stem. Wood, soft, light, tough, and close-grained, of agreeable fragrance; good for joiners and cabinet work ; often in request for coach building.

APOCYNEÆ.

190. *Alstonia constricta*, F.M. (Fever-bark.) —Diameter 6 to 20 inches; height, 40 to 50 feet. This tree is of frequent occurrence in low moist scrubs, as well as in the dry brigalow scrubs. Bark thick yellow, deeply fissurated, of intense bitterness. It possesses the same properties as quinine, and is coming much into demand. One firm of wholesale druggists have had a contract to be supplied with eight tons of the bark.

SCROPHULARINEÆ.

190A. *Duboisia myoporoides*, R. Br.—Diameter, 3 to 10 inches; height, 12 to 30 feet. A tall shrub or small tree, of frequent occurrence on the banks of fresh and brackish water creeks in the neighbourhood of Brisbane. A decoction of the leaves is said to cure ophthalmic affections, and at present is much sought after by some of our wholesale druggists to export to other countries.

LEGUMINOSÆ.

191. *Acacia falcata*, Willd.—Diameter, 6 to 12 inches; height, 20 to 30 feet. Small tree ; wood hard, and much prized for making stockwhip handles.

192. *Acacia glaucescens*, Willd.—Diameter, 12 to 18 inches; height, 30 to 45 feet. Of frequent occurrence both in the scrubs and in open forest lands, wood close-grained, and prettily marked.

193. *Acacia Cunninghamii*, Hook.— Diameter, 9 to 12 inches; height, 20 to 30 feet. Small-sized tree, wood close-grained, and takes a good polish ; found on the banks of the Brisbane.

194. *Acacia salicina*, Lindl.—Diameter, 6 to 12 inches ; height, 30 to 40 feet. Found on scrubby land in the Darling Downs district ; timber close-grained, and nicely marked.

195. *Acacia implexa*, Benth.—Diameter, 12 to 16 inches ; height, 30 to 40 feet. Small tree found on open forest ground ; wood hard and close-grained.

196. *Acacia harpophylla*, F.M.—Diameter, 12 to 20 inches; height, 40 to 70 feet. A tall erect tree, of frequent occurrence in the brigalow scrubs; timber hard, heavy, and elastic, of a reddish colour; suitable for cabinet work.

197. *Acacia excelsa*, Benth. (Brigalow).—Diameter, 20 to 30 inches ; height, 50 to 80 feet. This species covers immense tracks of rich scrub land ; wood hard, close-grained, of a dark colour, giving a strong odour of violets.

198. *Acacia doratoxylon*, A. Cunn.—Diameter, 6 to 12 inches ; height, 20 to 35 feet. In scrubs and open forest ground, wood hard, and close-grained.

199. *Acacia decurrens* (*var. mollis*), Lindl.—Diameter, 6 to 10 inches ; height, 30 to 40 feet. This species is of very frequent occurrence through the Darling Downs. Bark used for tanning purposes.

200 *Acacia pendula*, A. Cunn. (Weeping Myall).—Diameter, 6 to 12 inches ; height, 20 to 35 feet. Small weeping tree, well known for its violet-scented wood, which is hard, close-grained, and beautifully marked ; used by cabinet makers and turners ; in high repute for tobacco pipes.

201. *Acacia stenophylla*, A. Cunn. (Ironwood).—Diameter, 15 to 24 inches; height, 40 to 60 feet. On open forest ground on the Darling Downs. Timber is very hard, heavy, close-grained, dark, beautifully marked, and takes a fine polish.

202. *Acacia neriifolia*, A. Cunn.—Diameter, 6 to 18 inches; height, 40 to 50 feet. Found on the Albert River; the trunk is beautifully streaked with green and white. The duramen is of a light yellow colour, not unlike yellow-wood, but somewhat harder.

203. *Acacia penninervis*, Sieb.—Diameter, 2 to 10 inches; height 6 to 12 feet. Scattered through open stony ridges. Bark used for tanning purposes.

204. *Acacia macradenia*, Benth. (Myall or Tony).—Diameter, 2 to 12 inches; height, 30 to 50 feet. In scrubs and open forests; beautiful black hard and close-grained wood, taking a very high polish.

205. *Acacia farnesiana*, Willd. (Dead Finish).—Diameter, 3 to 6 inches; height, 12 to 18 feet. Wood close, heavy, taking a good polish.

206. *Acacia cultriformis*, A. Cunn.—Diameter, 2 to 3 inches; height, 12 to 18 feet. Wood soft, nicely marked.

207. *Acacia linifolia*, Willd.—Diameter, 4 to 6 inches; height, 12 to 18 feet. In more or less open forests; wood soft and elastic, suitable for axe handles.

208. *Acacia fasciculifera*, F. M.—Diameter, 6 to 15 inches; height, 20 to 30 feet. Timber very hard, heavy, tough, and close-grained.

209. *Acacia Gnidium*, Benth.—Diameter, 6 to 12 inches; height 12 to 25 feet. Wood close-grained, black, hard, and takes a good polish.

210. *Acacia Bidwilli*, Benth.—Diameter, 10 to 16 inches; height, 20 to 30 feet. A small-sized tree; timber hard, close grained, and takes a good polish.

210A. *Acacia grandis*, W. H. (Bitter-wood Acacia). — Diameter, 24 to 36 inches; height, 40 to 60 feet. This large tree is found in the thickly wooded scrubs near Murphy's Creek. Bark thick, fissurated, and containing a good deal of tanning matter. The wood is hard; bitter to the taste when newly cut; beautifully marked, and suitable for cabinet-work and other purposes.

211. *Albizzia canescens*, Benth.—Diameter, 15 to 20 inches; height, 30 to 50 feet. A beautiful spreading tree, found in open forests; wood close grained and tough.

212. *Albizzia procera*, Benth.—Diameter, 18 to 24 inches; height, 30 to 60 feet. A very ornamental tree; timber is close-grained, easily worked, and in use for building purposes.

213. *Albizzia Thozetiana*, F.M.—Diameter, 12 to 30 inches; height, 40 to 60 feet. A tree of common occurrence in stony scrubs in the Kennedy District. Timber very hard, heavy, tough, and close-grained. May prove useful for gun stocks, &c., &c.

214. *Peltophorum ferrugineum*, Benth.—Diameter, 24 to 30 inches; height, 40 to 80 feet. A large and handsome tree, in the open forest ground in the Mackay district; the timber is much in request for cabinet-work, &c., &c.

215. *Erythrina vespertilio*, Benth. (Coral Tree).—Diameter, 12 to 15 inches, height, 30 to 40 feet. Rather frequent both on the

coast and in the interior; a beautiful tree when in flower; wood soft, and is used by the aborigines for making their shields.

216. *Castanospermum australe*, A. Cunn. (Moreton Bay Chestnut).—Diameter, 24 to 36 inches; height, 80 to 90 feet. A magnificent ornamental tree, with large pinnate, green glossy leaves; of frequent occurrence on the banks of rivers, &c. The timber is dark and prettily grained, not unlike walnut; occasionally used for cabinet work, for which purpose it seems to be well suited.

217. *Barkleya syringifolia*, F.M.—Diameter, 12 to 18 feet; height, 40 to 60 feet. A handsome and beautiful tree, with dense bright-green glossy foliage; on fertile banks and flats of rivers; also on basaltic ridges.

218. *Cassia Brewsteri*, F.M.—Diameter, 3 to 6 inches; height, 30 to 50 feet. A small tree, of frequent occurrence in the Brigalow scrubs; wood fine-grained.

219. *Daviesia arborea*, W.H. (Queen-wood).—Diameter, 6 to 12 inches; height, 15 to 30 feet. A very ornamental spreading, drooping tree, with bright green foliage; occurring upon the Darlington Ranges. Its wood is hard, close-grained, with beautiful pink streaked lines, and takes a beautiful polish. It is destined to take a prominent position with cabinet-makers, also for decorative plantations.

220. *Galactia teniuflora*, Willd.—Diameter, 3 to 9 inches; height, 200 to 300 feet. An immense woody climber, outgrowing the tallest trees in the moist scrubs; the trunk is often found coiled on the ground like a huge serpent; the wood is dark-brown, soft, spongy, and pervaded with numerous capillary tubes; apparently of little durability.

ROSACEÆ.

221. *Parinarium nonda*, F.M. (The Nonda tree of North-east Australia).—Diameter, 18 to 24 inches; height, 50 to 60 feet. A very ornamental spreading tree, with pendent foliage. The aborigines use the esculent drupes as food. Timber is soft, close-grained, and easily worked.

COMBRETACEÆ.

222. *Terminalia discolor*, F.M.—Diameter, 3 to 6 inches; height, 10 to 15 feet. Very plentiful in the scrubs near Rockhampton; wood close-grained and tough.

223. *Terminalia melanocarpa*, F.M.—Diameter, 6 to 12 inches; height, 15 to 25 feet. Occurring in the forest land near Port Denison; timber hard and tough, splitting freely.

MALVACEÆ.

224. *Lagunaria Patersoni*, Don. — Diameter, 18 to 30 inches; height, 40 to 60 feet. Found on the alluvial river banks of the Don River, Port Denison; timber white, close-grained, easily worked, and used for building purposes.

225. *Hibiscus tiliaceus*, Linn.—Diameter, 6 to 8 inches; height, 20 to 50 feet. Of frequent occurrence on the coast; wood soft; the bark is used for nets and fishing lines by the aborigines.

EPACRIDEÆ.

226. *Trochocarpa laurina*, R.Br.—Diameter, 6 to 12 inches; height, 20 to 30 feet. Very plentiful in the cedar scrubs; also upon the sandy ridges upon Frazer's and other islands. Wood hard, close-grained, tough; useful for turning, &c.

227. *Leucopogon lanceolatus*, R. Br.—Diameter, 3 to 6 inches; height, 12 to 15 feet. Very abundant on Stradbroke and Frazer's Islands; wood hard, close-grained, and beautifully marked.

CASUARINEÆ.

228. *Casuarina fibrosa*, W. H. (The Stringy-bark Forest Oak.—Diameter, 6 to 10 inches; height, 25 to 40 feet. A small-sized tree, occurring on ridges in the Darling Downs and West Moreton districts. Wood hard, nicely marked, and suitable for cabinet work.

RUTACEÆ.

229. *Citrus Cataphracta*, W. H. (The Finger-fruited Lime.)—Diameter, 3 to 4 inches; height, 12 to 20 feet. A handsome, small-sized tree or shrub with a clear stem of 3 to 6 feet, thickly covered with diffused branches armed with ascending small straight spines about a quarter of an inch long. The fruit is about three inches long and one inch in diameter; a very agreeable beverage is produced from its acid juice. The wood is hard, close-grained, and of a straw colour.

JAMES C. BEAL, Government Printer, William Street, Brisbane.